Copyright © 2000 by Constance Korosec & Leslie Piña
Library of Congress Catalog Card Number: 99-66077

All rights reserved. No part of this work may be reproduced or used in any form or by any means—
graphic, electronic, or mechanical, including photocopying or information storage and retrieval
systems—without written permission from the copyright holder.
"Schiffer," "Schiffer Publishing Ltd. & Design," and the "Design of pen and ink well" are registered
trademarks of Schiffer Publishing Ltd.

Designed by Leslie Piña
Layout by Bonnie M. Hensley
Type set in Lithograph/Korinna BT

ISBN: 0-7643-1030-5
Printed in China

Published by Schiffer Publishing Ltd.
4880 Lower Valley Road
Atglen, PA 19310
Phone: (610) 593-1777; Fax: (610) 593-2002
E-mail: Schifferbk@aol.com
Please visit our website catalog at **www.schifferbooks.com**

In Europe, Schiffer books are distributed by Bushwood Books
6 Marksbury Avenue Kew Gardens
Surrey TW9 4JF England
Phone: 44 (0)208-392-8585; Fax: 44 (0)208-392-9876
E-mail: Bushwd@aol.com

This book may be purchased from the publisher.
Include $3.95 for shipping. Please try your bookstore first.
We are interested in hearing from authors with book ideas on related subjects.
You may write for a free printed catalog.

Contents

Acknowledgments
—4—

Introduction
—5—

Chapter 1: Solids, Textures, & Novelty
—6—

Chapter 2: Stripes & Plaids
—18—

Chapter 3: Geometric
—32—

Chapter 4: Abstract
—46—

Chapter 5: Floral
—69—

Chapter 6: Batik & Paisley
—86—

Chapter 7: Pictorial
—94—

Glossary
—108—

Selected Bibliography
—111—

Index
—112—

Acknowledgments

We would like to express our thanks to the following who so generously provided fabrics for us to photograph for this book: American Fabrics, Lois Epstein, Shirley Friedland, Knoll Textiles, Irene Sladewski, the Ursuline College Resource Library, and several anonymous sources. Thanks again to Ramón for his indispensable assistance with the photography; to the Ursuline College library staff; and to Peter Schiffer, Jennifer Lindbeck, and the gang at Schiffer publishing.

Introduction

Fabric of the 1970s was as distinct as the clothing into which it was made. Bold colors and patterns that dominated fabric design ranged from flamboyant to garish. Dizzying complementary color and pattern combinations, called psychedelic, were adapted from sources as diverse as non-Western folk art and Western technology. Some were inspired designs; some were arbitrary and thoughtless copies of inspired designs. Both represented the decade.

There were also two different categories of fibers used in the 1970s—the naturals and the synthetics. There were wonderful natural fabrics in silk, cotton, rayon, flax and wool that represented the decade. The '60s counterculture advocated things like "earth shoes," native folk art, a relaxed and/or rustic life style, natural foods, and natural fibers. When their radical clothing styles and colors became more familiar and ultimately more fashionable with the general population, synthetic stretch knits became the rage and coexisted with the natural fabrics.

The companion volume to this book, *The Synthetic '70s: Fabric of the Decade,* focuses on the synthetics. Extreme fashion and wildly printed fabrics inspired by counter-cultural ideals ironically were made from synthetics at least as often as the natural fibers for which they were intended.

The following photographs (mostly close-ups) of fabric swatches cannot be felt. However, they certainly provide a sense for the visual delight the actual fabrics convey. Each fabric swatch acts as a miniature historical document that can be dated to a specific period and can serve as a reference for future designs. The glossary found in the back of the book focuses on terms associated with and pertaining to natural and man-made fibers. These terms and dates will aid one in understanding some of the materials, processes, and distinctive qualities that these fabrics possess. The bibliography will lead the reader to more detailed technical, historical, or otherwise additional information. As with the companion volume, *The Synthetic '70s: Fabric of the Decade,* our main intent here is to present a visual sampling of some naturally '70s morsels for the shear pleasure of seeing them.

Chapter 1

Solids, Textures, & Novelty

Apple green corduroy, light, soft, comfortable, and wrinkle resistant for relaxed living.
Crompton
1975

Beige loose plain weave.
Early 1970s

Magenta brushed mohair.
Anglo
1972

Raspberry sherbet cotton chambray permanent pleating.
Galey & Lord
1970

Rich yellow, gold corduroy all-purpose fabric.
Cone
1970

Coral sculptured rib surface, the ultimate in corduroy.
Cone
1970

Top left: Brown carpetbagger chenille tweed with warm hand-loomed look.
Cohama
1970

Bottom left: Tweedy, rugged woven fabric great for men's sport jacket.
Cohama
1971

Top right: Soft silver gray Panne brocade warp knit with an argyle look.
Wamsutta Fabrics
1970

Bottom right: Brown tweed-like pattern effect, made in Vallemosso, Italy.
Zegna
1971

Top left: A subtle, soft reddish color knit blend of wool and flax created by Dixie Yarns.
Bronzini Sportswear
1970

Bottom left: Texture is key for a soft, subtle misty color blend creating a hand knit look.
Held
1977

Top right: Raschel knit construction Shetland.
Novelty Textiles
1972

Bottom right: Texture stripe on solid, single knit cotton for men's and women's sportswear.
Caro-Knit
1973

Shetland double knit with two-way stretch.
Bravo Knits
1972

Beige with turquoise stripe Tussah silk tweed for sportswear.
Onondaga
1973

100% wool multi-color yarn fabric used for suiting and coating.
Early 1970s

Flame retardant upholstery cloth of 100% rayon.
Orinoka Mills
1972

Blue and white fine wale corduroy which looks like sweater knit 100% cotton.
Cone Mills
1974

Yellow bulky sweater knit, lofty thick fabric.
Held Fabrics
1974

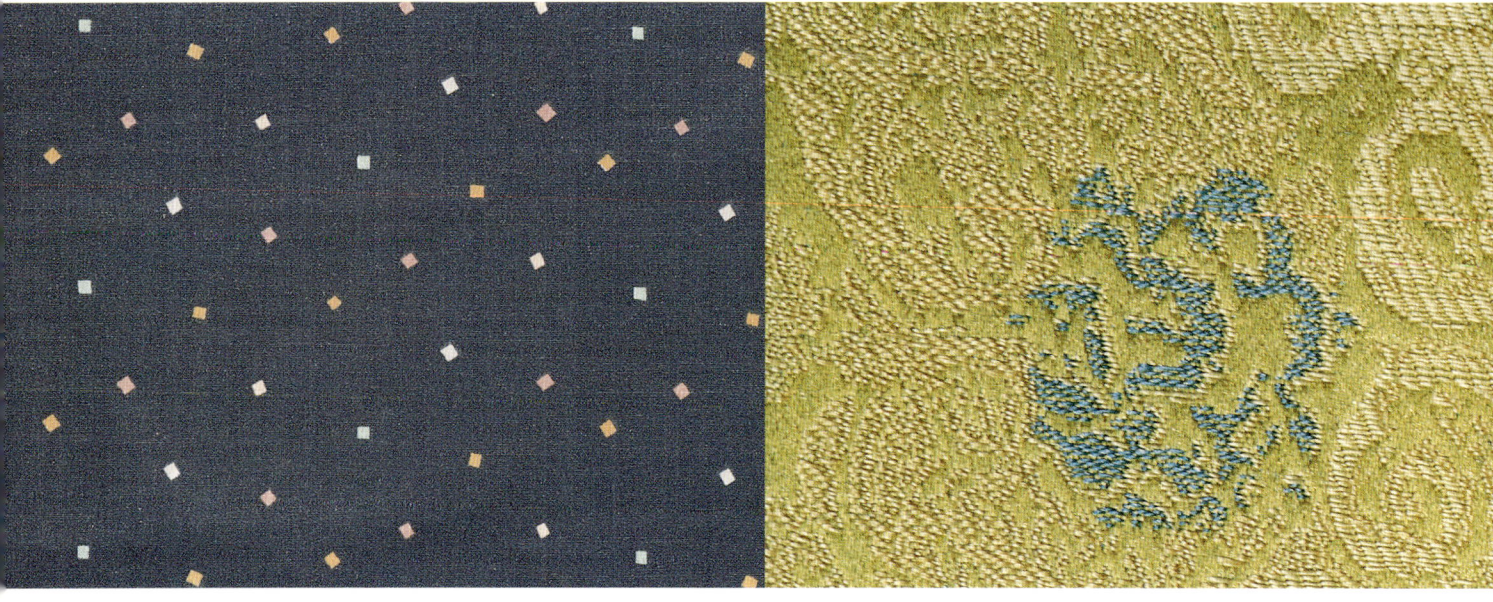

Cotton polyester "little squares" print.
Early 1970s

Korean raw silk brocade for interior furnishings.
Silkorea
1971

Denim tie-dye.
Mid 1970s

Earthtone upholstery velvet.
Collins & Aikman
1973

White cotton eyelash novelty fabric.
Mid 1970s

Cotton eyelash fabric, reversible pink butterfly.
Mid 1970s

Dark green/yellow cotton eyelash fabric.
Mid 1970s

Dark green/yellow cotton eyelash fabric,
reverse side.
Mid 1970s

Open, loose weave with cotton thread, pulled through the fabric.
Mid 1970s

Cotton blend with embroidered eyelet.
Mid 1970s

Woven cotton, a post-folkloric look for dresses and blouses for women's wear.
Earl-Glo Fabrics
1973

Wine, brown, and white cotton seersucker stripe blend.
Russell Mills
1971

Loose weave in flax natural color, used for casements to allow maximum light penetration, without loss of privacy.
The Belgian Linen Association
1976

Neutral translucent textured cotton and flax blend used for casement, great for curtains and drapery.
Sunrise Drapery Co.
1977

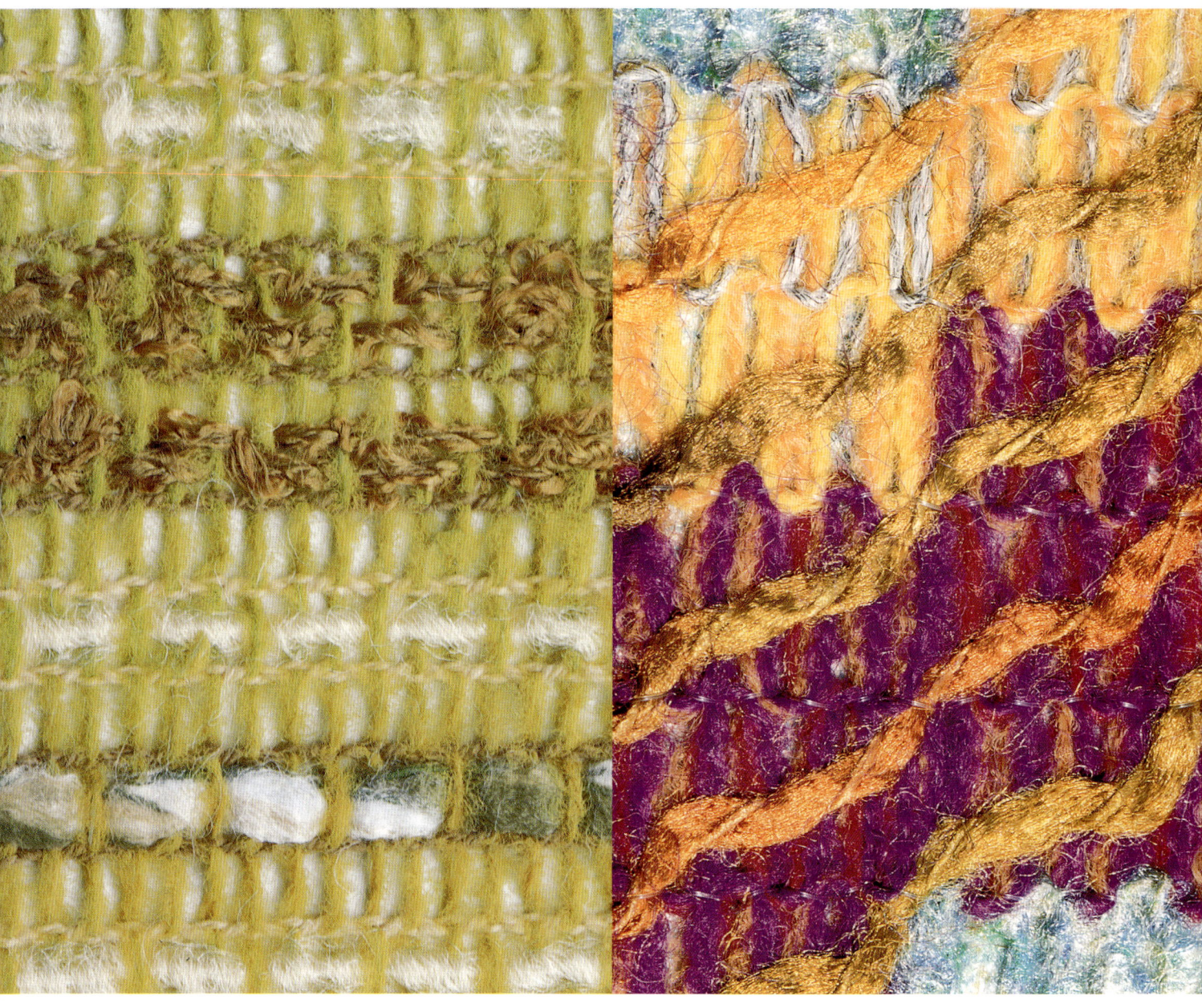

Drapery, weightless, flax open weave, linear, embroidery, and lace effect.
Cannon Mills
1974

Peasant mohair and wool embroidered look in knit, bold colors.
Lowenstein
1972

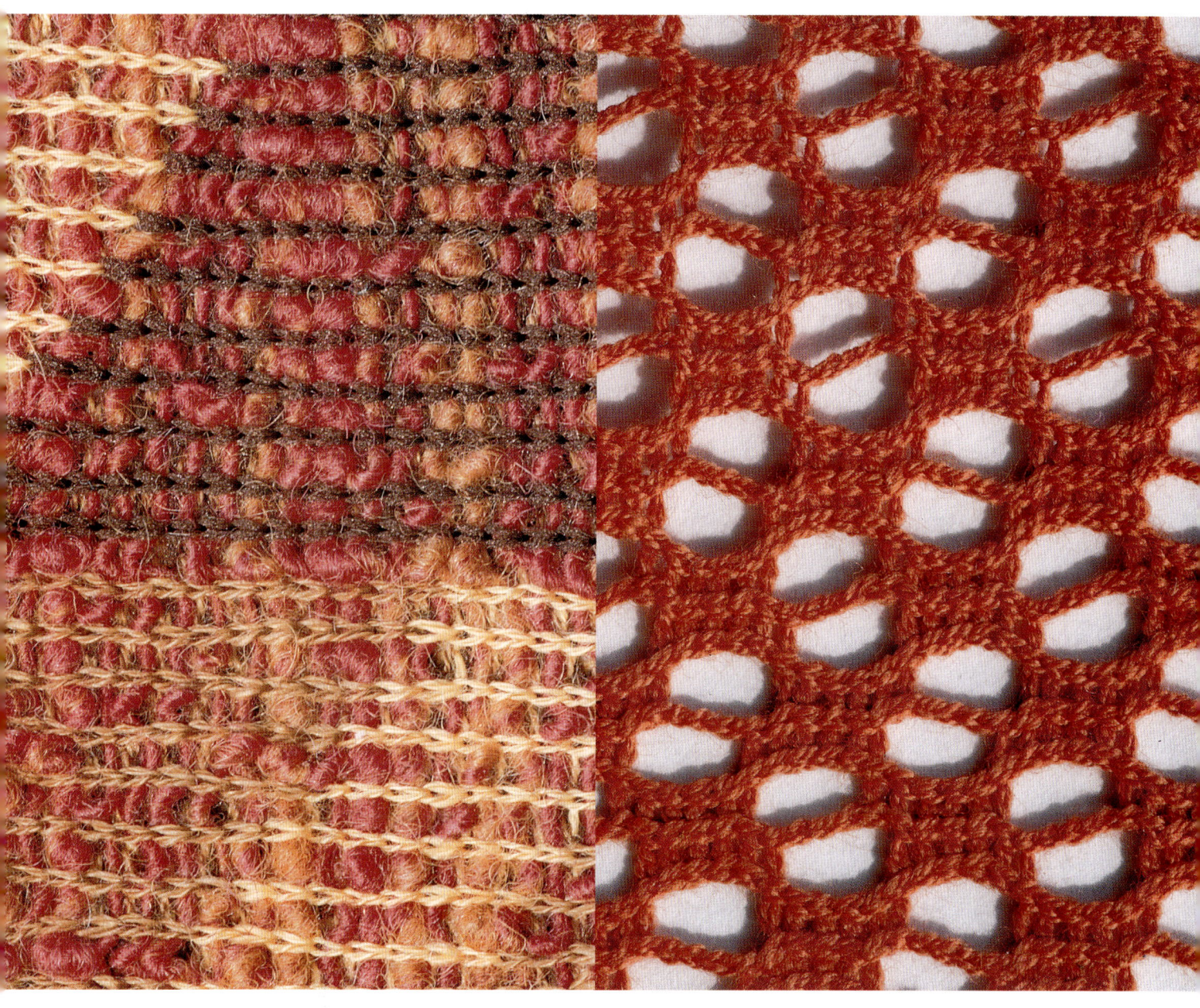

Textured, upholstery, cotton, and flax double knit for furniture.
Sulzer Bros.
1974

Red/orange string cotton knit used for vests and tunics.
Wedgwood
1970

CHAPTER 2
STRIPES & PLAIDS

For men who want to be fashion leaders, bold stripes in cotton shirting.
Dan River
1970

Brilliant stripes in men's cotton shirting.
Dan River
1970

For the bold new look in men's cotton shirting, superb coloring.
Dan River
1970

Pink and blue striped men's cotton shirting.
1970

White, green, red, blue, yellow, and lavender cotton stripe on white ground.
Mid 1970s

Bold wide stripe for bedding home fashions.
Springmaid
1978

Gray flannel wool with white and black acrylic stripes.
Stevens Hockanum
1970

Dull red striped cotton gauze.
Early 1970s

Dobby design seersucker blend in
browns and blues.
Russell Mills
1972

Un-denim denims 50% Kodel polyester/50% cotton; men
and women wore pants of exactly the same fabric.
Swift
1970

Cotton rib knit with elasticity for women's, men's,
and children's wear.
Stevcoknit
1973

Striped cotton blend knit.
Early 1970s

"Boutique" un-denims rich wide stripes, narrow stripes, uneven stripes, and all cotton fabric.   Swift   1970

Trompe l'oeil red and white woven illusion print of cotton and polyester.   Ameritex   1973

T-shirt single knit fabric for men's sport shirts in pastel colors. Caro-Knit Inc.   1973

Pastel rainbow stripes, cotton. Early 1970s

Bleeding of pattern is ikat-inspired, suitable for home furnishing.   Orinoka   1978

Navy, red, and white stripe, first warp knit terry in a sculptured design of combed cotton and Phillips 66 stretch nylon.
Domestic Fabrics Corp.
1971

Striped stretch terry cloth sold in one-stop centers by Singer.
1971

Green, pink, orange, and black cotton and polyester blend, striped terry cloth.
Mid 1970s

Gucci's two color stripe found on shoes, bags, and wallets.
Jack Rose New York Trim n' Fashion
1975

Korean raw silk for women and men's apparel, made in Korea.
Silkorea
1971

Cotton polyester blended stretch velour in bold red and blue stripe.
Early 1970s

Top left: Cotton polyester print stripe in reds and blues.
Early 1970s

Top right: Multi-color print stripe on white background woven cotton.
Early 1970s

Bottom left: Loosely woven blue and earthtone cotton print.
Early 1970s

Bottom right: Cotton print made with rows and rows of borders which can be cut apart and applied creatively.
Early 1970s

Top left: Wide and narrow print stripes, blue, earthtone, and white cotton.
Early 1970s

Bottom left: Bold multi-color cotton borders.
Early 1970s

Top right: Wide and narrow print stripes, blue earthtone, and white cotton.
Early 1970s

Bottom right: Bold cotton borders in green, earthtone, and white.
Early 1970s

Black background with soft, muted stripes, cotton.
Mid 1970s

Striped silky, light weight surah, clinging cotton and silk, wrinkle resistant and washable.
Ameritex
1970

Double knit bias plaid knitted diamond pattern of polyester and cotton blend, used in skirts and vests.
Galey & Lord
1972

Fashion un-denim blend of 50% Kodel polyester/ 50% cotton striking herringbone for casual wear.
Swift
1970

All wool stretch Shetland wool designed for men's sport jackets, suits, and slacks.
J. P. Stevens & Co.
1972

New sport shirt look in beautiful herringbone soft color for men.
Dan River
1970

Blend 50% Kodel polyester /50% cotton striking plaid for fashion un-denim.
Swift
1970

Camel, gray, and white wool.
Mid 1970s

Kodel cotton blend denim classic plaid for the "un-denim generation" created by Swift.
Swift Textiles
1970

Wool with the look of linen in muted tones.
Mid 1970s

Printed linen plaid for sportswear.
Avondale
1973

Cotton blend wash and wear fabric in primary color plaid.
Dan River
1972

Midnight blue, soft apricot beige, and blue gray, "Cabana Stripes." Cotton glazed chintz for Caribbean casual decoration.
Waverly Fabrics
1979

Fashion stretch stripes and plaid for cotton pants.
Swift
1970

Blend 50% Kodel polyester/ 50% cotton look of wool for the un-denim fashion in red, blue, and gold plaid.
Swift
1970

Brushed plaid great for outerwear in multi-color wool blend.
Tromer Fabrics
1973

All wool gray, rust, white, and gold plaid.
Mid 1970s

Navy and white plaid cotton gauze.
Mid 1970s

Bold checks seersucker in pastel lavender and blue, a nostalgic link with the past.
Cohama
1972

Chapter 3
Geometric

Black and white harlequin pattern, cotton blend.
Early 1970s

Cream, black, and brown geometric double knit used in women's coats.
Early 1970s

Opposite page:
Top: Colorful soft cotton in stripe/plaid.
Late 1970s

Bottom: "Éclat" a Anni Albers design, 50% cotton/50% mohair, moth-proof used for upholstery. Shown in five colorways.
Knoll Textiles
1976

Lime and turquoise bold geometric wool knit.
Late 1970s

Elaskeen soft neutral color wool and silk. Early 1970s

Elaskeen soft color wool and silk. Early 1970s

Black, beige, and white, heavy cotton in bold geometric pattern. Mid 1970s

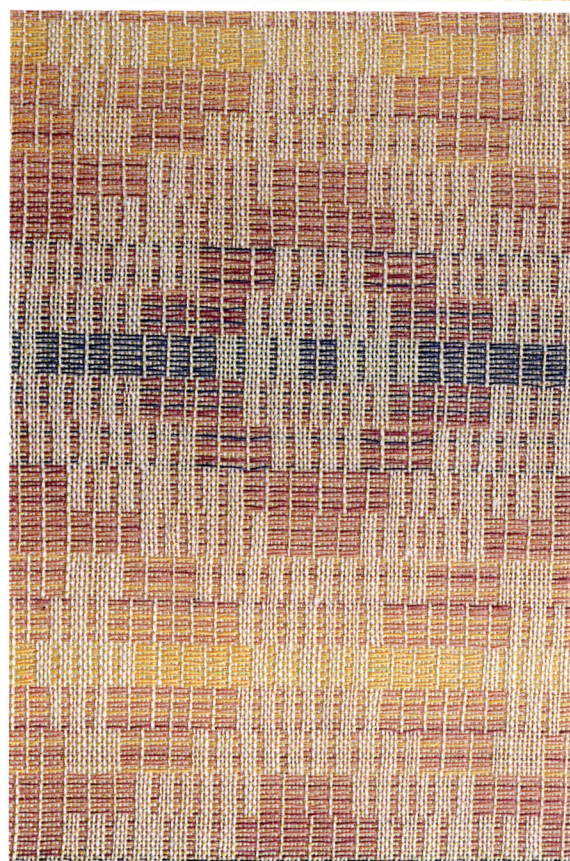

Top left: Red geometric, single knit cotton for men's and women's sportswear.
Caro-Knit   1973

Botttom left: A woven tapestry look in men's shirting from an American Indian inspired pattern.
Dan River   1970

Top right: Middle Eastern inspired geometric of varying tones of neutrals, gold, beige, and gray combinations.
Riverdale Drapery Fabrics   1977

Bottom right: Printed and sueded cotton, African inspired, made in Holland.
Hatéma-Texoprint   1971

Top left: Bright gauzy ethnic print with flavor of Mexico or Peru.
Wamsutta
1974

Bottom left: Black, white, and yellow colorful wool knit.
Late 1970s

Top right: Blue and turquoise print, made in Holland.
Hatéma-Texoprint
1971

Bottom right: Aztec Indian inspired, woven cotton.
Early 1970s

Top left: Mini geometric and small foulard print patterns on velveteen.
Crompton
1979

Top right: Blended knit for sportswear, primary colors.
Dan River
1974

Bottom left: Cotton blend print single knit sportswear fabric.
Cyanamid
1974

Bottom right: Red background with black honeycomb design silk.
Late 1970s

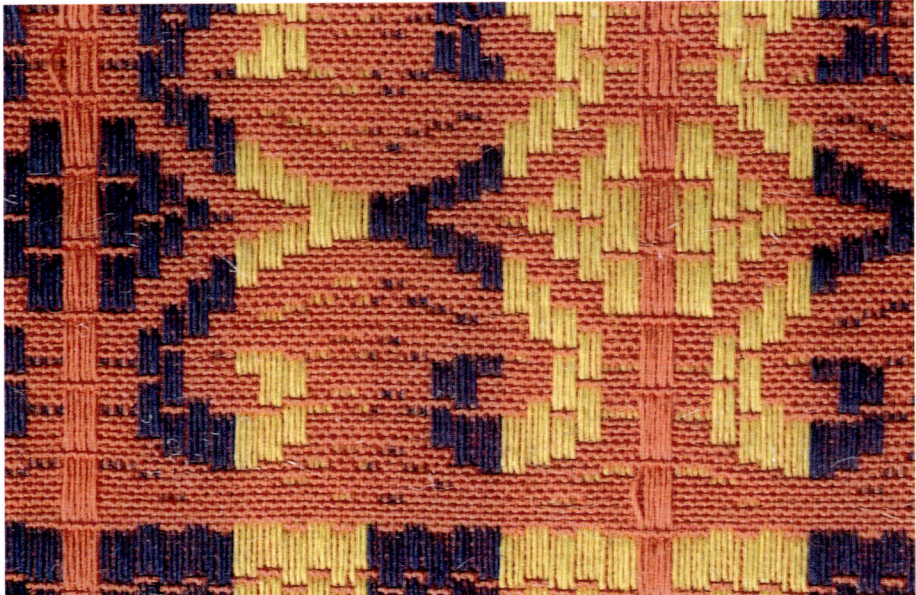

Top left: Tapestry woven novelty fabric cotton blend inspired by ancient cultures.
Dan River Mills
1970

Top right: 100% cotton knit, blue and tan pattern.
Galey & Lord
1970

Bottom left: Denim BouBoudima indigo print (name for exotic prints), made in Helmond, Holland.
Vlisco
1974

Bottom right: A polyester spun acetate multi-color novelty.
C & V Fabrics
1971

Top left: Men's cotton shirting, Renaissance pattern in earthtones.
Dan River
1970

Bottom left: Men's wear printed cotton for men's ties.
R.B.S.
1974

Top right: Red, white, and blue beautiful geometric bold stripe used in pants.
Cone Mills
1970

Bottom right: Woven to achieve the look of crewel embroidery in textile design.
Dan River
1975

Opposite page: Hot color diagonal shapes for bedding home fashions.
Springmaid
1978

100% cotton knit, white on purple inspired by Japanese rice paper designs with an oriental look.
M. Lowenstein & Sons
1970

Woven in intriguing pattern of purple on white.
Soptra
1970

Opposite page:
Top: Multi-color dots on white, cotton.
F.C.I.
Early 1970s

Bottom: Dots return with fashion pique voile in a neutral, bold circle pattern.
Spring Mills
1970

Opposite page:
Top: Exotic cotton print made in Holland, deep, rich earthtones.
Hatéma-Texoprint
1971

Bottom: Bold fabric double woven reversible jacquard popular in women's coats.
Deering Milliken
1970

Two-tone pink cotton blend bull's-eye.
Deering Milliken
1970

Jacquard Belgian linen interlinking arches, reminiscent of Art Deco period and used for drapery, upholstery, and wall covering.
The Belgian Linen Association
1976

100% cotton jacquard single knit in soft pastel colors.
1971

Soft pastel color, woven cotton.
Mid 1970s

45

Chapter 4
Abstract

Left: Woven cotton damask in natural neutrals, to be used in fashion and home interiors.
Orinoka
1978

Right: Two-faced, reversible wool fabric double weave.
Milliken
1972

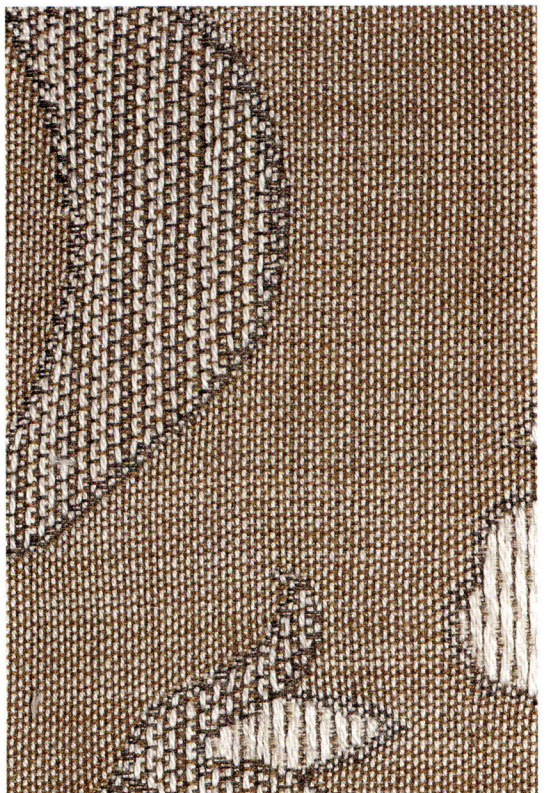

Soft muted pink, gold, and green silk print.
Mid 1970s

Opposite page: Boussac design in bright amber with raspberry linear motif, rayon.
Early 1970s

Japanese inspired, one color cloud cotton print.
Riverdale
1978

Black and white silk, bold vertical irregular stripe.
Early 1970s

Brown and red silk, wave stripe-like pattern.
Early 1970s

Top left; Batik inspired, midnight blue background, cotton, made in Holland.
Hatéma-Texoprint   1971

Bottom left: Indigo blue cotton print by Vlisco of Holland.
Continella Industries   1973

Top right: Indigo block print by Vlisco of Holland, cotton, African inspired.
Continella Textile Corp.   1973

Bottom right: Denim BouBoudima indigo print (name for exotic print), made in Helmond, Holland.
Vlisco   1974

Bold colorful rayon.
Mid 1970s

Raspberry and other bold colors on black, rayon.
Mid 1970s

Minimalist pattern in primary colors on heavy cotton duck cloth.
Late 1970s

Yellow, orange, blue, and purple on cotton.
Early 1970s

Opposite page: Top left: Pink, orange, aqua, and green on cotton.
Early 1970s

Top right: Black background with rose and white design on silk.
Mid 1970s

Bottom left: Chintz colorful brights, light up dark background, stylized floral.   Greeff   1977

Bottom right: 100% cotton duck with deep Persian coloring, used in sportswear.   West Point Pepperell   1970

Top left: Brown and rose tones with a hint of pale pink, loose weave cotton.
Early 1970s

Bottom left: Damask brocade effect, 100% cotton Zepel finish, floral contemporary.
Bloomcraft
1977

Top right: Shades of earthtones on woven cotton.
Mid 1970s

Bottom right: Multi-color Korean raw silk for interiors and apparel.
Silkorea
1971

White background with pink, orange, blue, and green colorful designs on cotton.
Early 1970s

Two-tone green cotton blend, raindrop argyle print.
Deering Milliken
1970

Reproduction of 2000-year-old silk design from People's Republic of China, stylized clouds and birds' wings.
China Trade Corporation
Seta-Sole International Ltd.
1975

Combination of soft, subtle shades on cotton contrasted with crackle batik effect. Inspired by the nostalgic look of ante-bellum wallpaper.
Spring Mills
1970

Large intricate pattern, beautiful cotton velvet, batik inspired motif for upholstery.
Schumacher
1978

Black, dull red, gold, and white on cotton.
Early 1970s

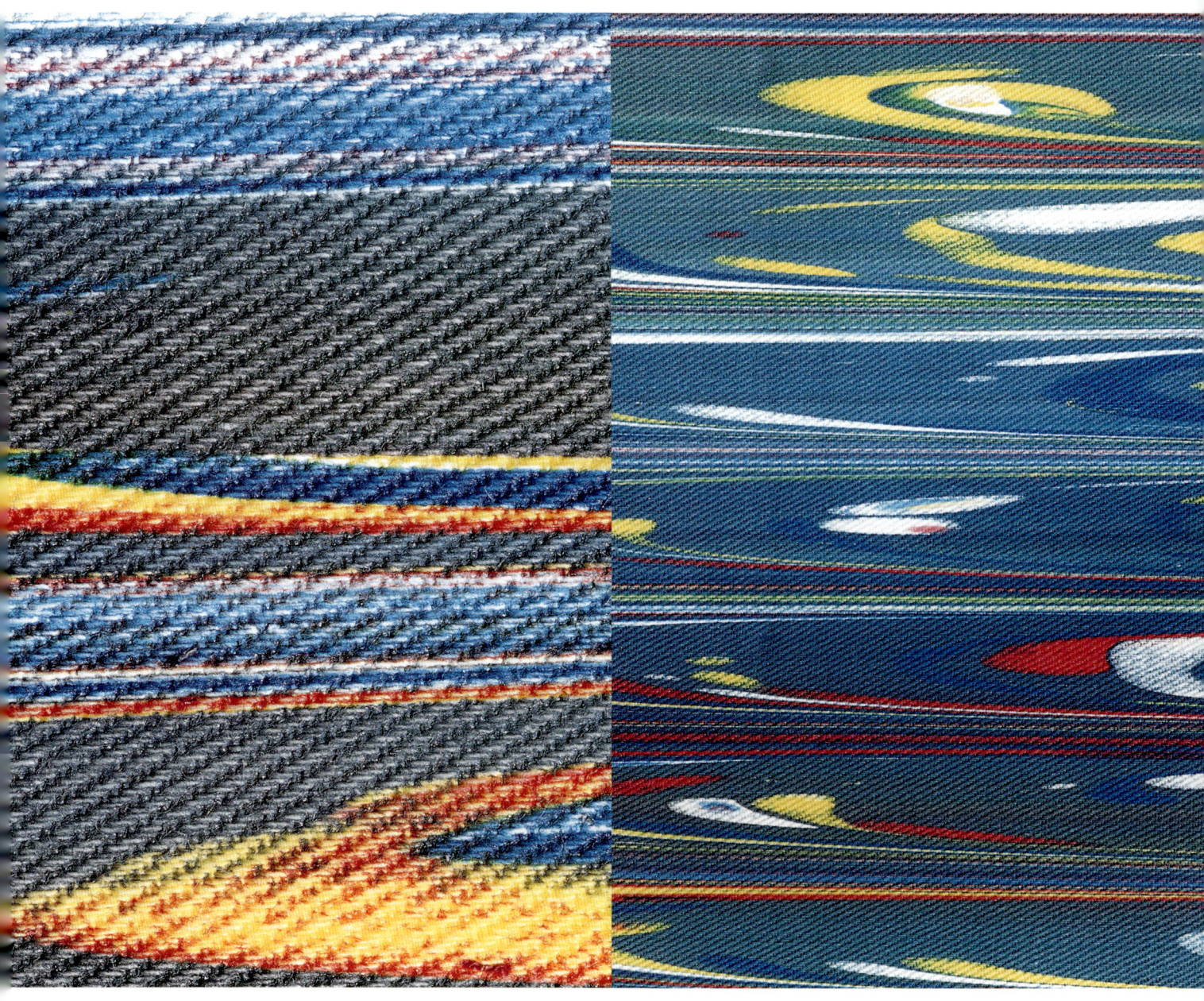

Close-up view of painted denim pants fabric with bold cathedral print, no two prints exactly alike.
Graniteville
1970

Painted denim pants with bold cathedral print, no two prints exactly alike, showing larger pattern.
Graniteville
1970

Opposite page: Blues and gold, in stylized seed pattern on rayon.
Early 1970s

White background with green, blue, and red complex pattern on rayon.
Early 1970s

Challis crepe, soft floral on a black background.
Cranston
1979

Bold colorful wool and synthetic blend. Mid 1970s

Cotton screen print, large repeat, permanent press.
Wamsutta
1973

Cotton print, associated with the Provence region of France, produced in hand blocks.
Crantex
1971

Challis rayon and wool combination in red, white, and varying blues.
Late 1970s

Different view.

Cotton voile in bold hot colors, paisley inspired pattern. Early 1970s

Red, white, green, yellow, black, and blue print on rayon.
Mid 1970s

Greens with black and white abstract on rayon. Mid 1970s

Multi-color soft watercolor effect on silk. Mid 1970s

Brilliant color flower garden floral knit.
Transprints International
1975

Watercolor effect on silk in blues, yellow, and pink.
Mid 1970s

White, gray, green, and red abstract on silk.
Mid 1970s

Cream background with black, brown, and aqua winter scene on very sheer cotton.
Early 1970s

Soft shades of summer colors on silk.
Early 1970s

Soft design, subdued earthtone colors, silk blend print.
Mantero
1978

Very pale sherbet colorful cotton print.
J.B.J. Fabrics Inc.
Mid 1970s

Chapter 5
Floral

Top left: Subdued elegant floral woven cotton print.
Covington
1978

Top right: Rich cloisonné-inspired design with exotic motif, cotton.
Covington
1978

Bottom left: Bold purple with yellow and pink print on cotton.
Early 1970s

Bottom right: Velveteen/corduroy cotton print, pink and red.
Crompton-Richmond
1970

Top left: Multi-color floral print cotton corduroy.
Cone
1970

Top right: Bright colored folded fabric design with small flowers, cotton blend.
Mid 1970s

Bottom left: Chintz with a crisp hand, 100% cotton, large floral.
Cyrus Clark
1977

Bottom right: Blue background with red and pink flowers on wool knit.
Mid 1970s

Opposite page:
Top: Large floral crepe, deep lush color, tactile hand.
Onondaga
1970

Bottom: Midnight blue with pink floral, rayon.
Early 1970s

Opposite page:
Top: Small floral print on challis. Early 1970s

Bottom: 100% cotton printed knit, made in Holland. Hatéma-Texoprint 1971

Top left: Soft supple floral on blue cotton corduroy for casual clothes.
Cone
1978

Bottom left: Brown and beige floral, silk.
Mid 1970s

Top right: Red with flecks of color on silk.
Mid 1970s

Bottom right: Beige background with red floral, cotton.
Mid 1970s

Top left: Patchwork motif on 100% cotton.
Mid 1970s

Bottom left: Brown, gold, and pink floral print on silk.
Late 1970s

Top right: Cotton print resembling batik.
Mid 1970s

Bottom right: Heavy cotton, bold floral on white background.
Mid 1970s

Top left: Muted color, light weight Victorian floral cotton knit.
Transprints International
1975

Top right: Flowering leafy motif, a Jacobean tapestry perfect for upholstery, cotton.
Greeff
1978

Bottom left: Chintz with earthy oriental-inspired print, a favorite of the home furnishing industry.
Everfast
1970

Bottom right: Pale soft neutral colors on silk.
Mid 1970s

White background with black sketchy floral, on silk. Mid 1970s

100% combed cotton velveteen with luxurious touch. Compton 1976

Cotton velveteen, blue and rust English cabbage roses on beige background. Mid 1970s

Cream background with green and floral print on cotton. Mid 1970s

Stylized leaves in warm earthtones on cotton. Late 1970s

Crepe challis soft floral on a wine background. Cranston 1979

Raspberry background with bold, stylized floral on rayon. Late 1970s

Chinese red with raspberry floral on rayon.
Late 1970s

White background with bold poppy motif on silk.
Late 1970s

100% cotton challis print, black background floral.
Ameritex
1971

Soft tapestry design floral knit.
Transprints International
1975

Large floral cluster on cotton chintz.
Stroheim & Romann
1977

Green silk floral.
Mid 1970s

Opposite page: Fruit and leaf motif in red, purple, and varying greens, cotton. Mid 1970s

Similar pattern with added turquoise. Mid 1970s

Hawaiian cotton print in bold colors. Mid 1970s

Bright pink with green and white on a black background, rayon.
Late 1970s

Opposite page: Jungle print in browns, red with a hint of rose color on cotton.
Late 1970s

Chapter 6
Batik & Paisley

Left: Denim BouBoudima indigo print, made in Helmond, Holland.
Vlisco
1974

Right: Denim BouBoudima indigo print, made in Helmond, Holland.
Vlisco
1974

Left: Cotton double knit paisley tear drop pattern on white background.
Galey & Lord
1972

Right: Contemporary cotton double knit paisley classic pattern on gray background.
J. P. Stevens
1975

Persian paisley pattern of colorful mini dots, cotton blend.
Wamsutta
1975

Cotton paisley print in reds.
Mid 1970s

Soft elegant velveteen fine paisley detail, 100% combed cotton for blouses, skirts, dresses, and jackets.
Crompton
1978

Printed muted tones, paisley type pattern cotton knit.
William Heller, Inc.
1971

Opposite page: Fine cotton with bold paisley motif in hot colors.
Late 1970s

Rayon in bright orange with white motif.
Late 1970s

Plush printed corduroy, fine wale cotton polyester blend, wrinkle resistant.
Crompton
1975

Cotton voile in summery colors.
Late 1970s

100% silk paisley print in hot colors on green background. Early 1970s

Batik-like print on cotton, cool colors.
Late 1970s

Muted colors, border design on cotton.
Late 1970s

Dutch batik hand-blocked print, slightly off register, which adds to the fabric's appearance.
Hatéma-Texoprint
1970

Polyester/cotton blend, aqua background with hot colors.
Late 1970s

Green and red challis, fashion knit print, cotton blend.
Wullschleger
1974

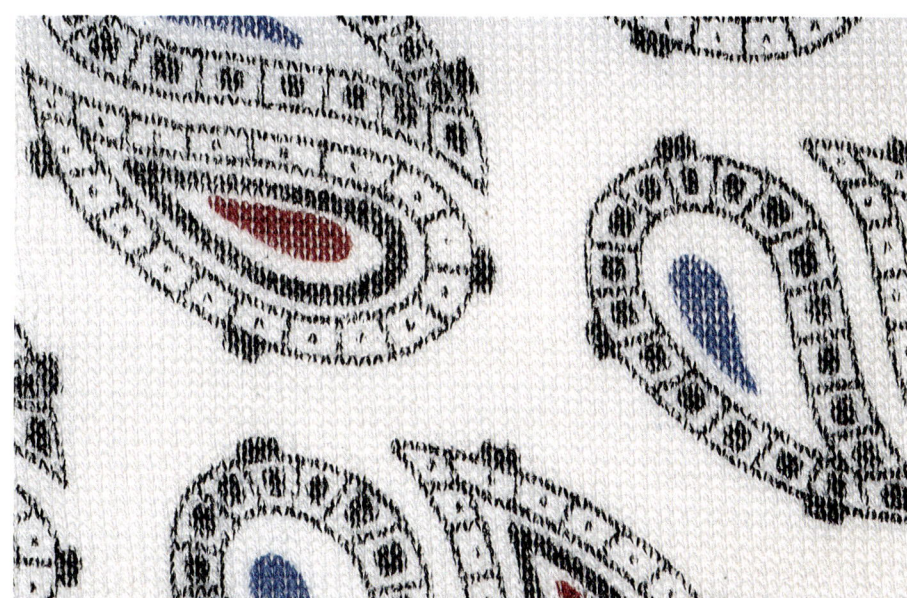

Paisley single cotton knit for men's sport shirts.
Quality Knitting Mills
1973

Browns and green, intricate pattern on cotton.
Schwartz-Liebman Textiles
Late 1970s

Chapter 7
Pictorial

Purple cotton single knit with a message of love.
Stevcoknit Fabric Company
1970

Cotton blend, navy with lettering.
Late 1970s

Opposite page:
Top: White cotton with navy newspaper printed design.
Late 1970s

Bottom: Different section of the newspaper.
Late 1970s

A WAGON INTENDED FOR GENERAL PURPOSES,
WEIGHS BUT 600 POUNDS.

Is finished ready to paint and trim. Inquire of your carriage-maker, or the sole manufacturers,
S. N. BROWN & CO., of Dayton, Ohio.

W. L. HAYDEN, Teacher of Guitar and Violin. Agent for Firth's CELEBRATED Italian Guitar & Violin Strings. Dealer in Musical Instruments, Music, and Sheet Music. Catalogues free. 167 Tremont St., Boston, Mass.

$475 A MONTH to Agents to sell our Staple articles. Samples free. C. M. LININGTON, Chicago.

$15 to $20 A DAY guaranteed using our Well Auger & Drill in good territory. Highly illustrated Catalogue Free. W. GILES, St. Louis, Mo.

$325 A MONTH and expenses to good canvassers. Address T. A. DEAN, New Bedford, Mass.

THE CHILDREN'S HOUR

The best magazine for children in the world.—N. Y. Times.

REED & BARTON,

MANUFACTURERS OF

the ELECTRO-PLATED WARE.

FANCY VESTS

6,000,000 Pairs of

SILVER TIPS

Are sold annually, the question is, Where are they worn? On the feet of course.

CHENEY BRO'S

American Gros-Grain Silks,

IN BLACK STRIPES,

AND ELEGANT FALL COLORS,

Of Shades especially adapted for Suitings.

$72.00 EACH WEEK.—Agents wanted every where. Business strictly legitimate. Particulars free. Address J. WORTH & CO., St. Louis, Mo.

Stretch fabric 100% cotton twill, Woodstock montage of images.
Avondale
1970

Cotton with patchwork-like pattern in primary colors.
Mid 1970s

Cotton with figural motif in muted colors.
Mid 1970s

Rayon with navy and white dot pattern and scattered figures.
Mid 1970s

100% cotton knit printed with bright rows of red and blue houses.
Alamac Knitting Mills
1970

Flannel blend for sleepwear, robes for infants, girls, boys, women, and men.
Lowenstein
1974

Navy with thin white pin stripes and pink elephants on cotton.
Mid 1970s

Opposite page: Detail.

Cotton gauze in white, with Art Nouveau revival tendrils, flowers, and female heads. Mid 1970s

Cotton blend knit with rust and white female faces. Mid 1970s

Cotton blend with tropical figural scene. Mid 1970s

Different section.

Cotton voile jungle print with leopards. Mid 1970s

Cotton voile aquarium print. Mid 1970s

Cotton knit, white background, dog wearing boots, designed by Oleg Cassini.
Mid 1970s

Cotton chintz traditional floral with a botanical look for the home furnishing industry.
Cyrus Clark
1978

Print with jungle theme, very bright, washable and crease resistant.
Crantex
1970

Oak leaves and acorns, cotton blend knit, off shade green and tan on black.
Couleur
1975

Cotton voile print with black and white stylized lion in jungle.
Early 1970s

Heavy cotton owl print on lime green background.
Mid 1970s

Cotton stylized geometric floral print used in shirting for men and women.
Kenmill
1972

Hawaiian cities print on cotton.
Early 1970s

Cotton challis, mythical unicorn in flower garden.
Early 1970s

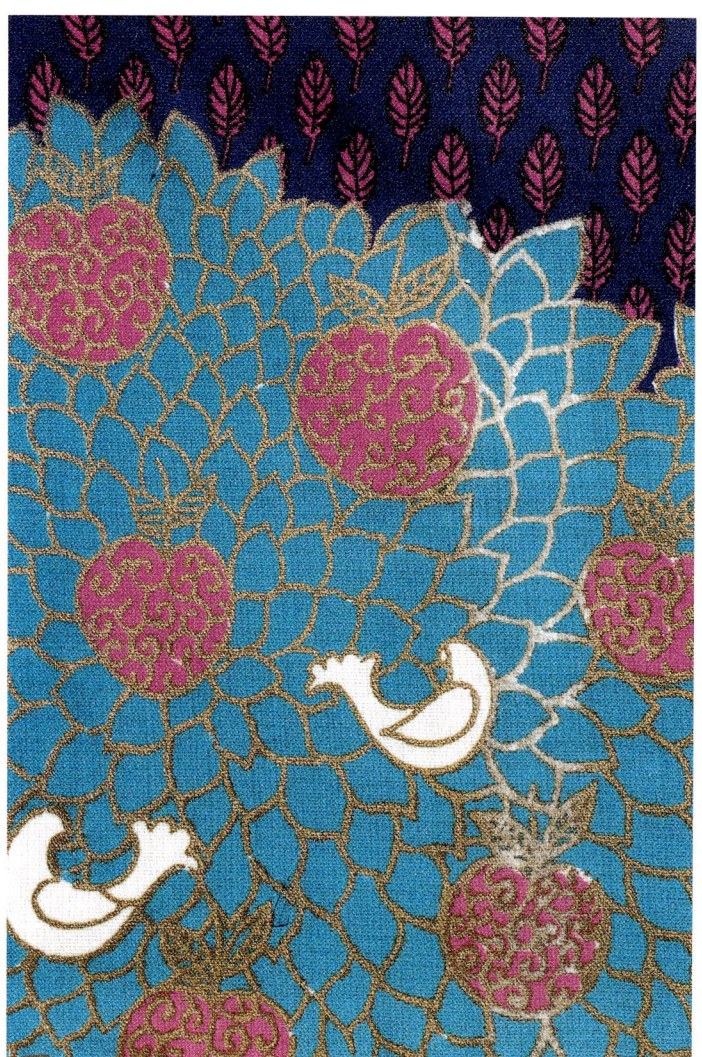

Cotton blend with two white doves, hot pink fruit, and aqua petals.
Mid 1970s

Detail of different section.

Opposite page: Hot bold colors on cotton.
Mid 1970s.

Glossary

Acetate - a manufactured fiber in which the fiber-forming substance is cellulose. (The look and feel often seems synthetic.)

Acetylation - chemical reaction in which cellulose is changed into cellulose acetate.

Air Jet - technique for bulking a filament yarn by subjecting it to jets of air under pressure.

Arnel - registered trademark of Celanese for its triacetate filament yarn.

Autoclave - a large chamber that can be sealed and filled with steam under pressure.

Avril - trademark of Avtex Fibers Inc. for their standard, high wet modulus rayon fiber having a round cross section.

Avril III - trademark of Avtex Fibers Inc. for their natural wet modulus rayon fiber having a multi-lobal cross section.

Barré - streaks that develop in a dyed knit fabric made with textured filament yarns.

Bi-component - a man made filament fiber in which two chemical components have been combined.

Bulking - the characteristic result of yarn texturing.

Cellulose - structural material found in plants like stems and seed pods.

Chemical Cellulose - a carbohydrate of complex molecular structure that forms the chief framework of plant cells and walls; the basic raw material of cellulosic fibers.

Coil - typical spring-like formation of a false twist stretch yarn.

Compacting - a technique for imparting texture to a finished fabric of thermoplastic fibers.

Coning - winding the textured yarn on a cone, ready for knitting or weaving.

Continuous Filament - a long continuous strand of a manufactured fiber; distinguished from all natural fibers (except raw silk), which are of short staple length.

Conventional Twist - the original method developed by Herberlein in Switzerland for making stretch yarns by twisting, setting, and untwisting in three stages.

Co-polymer - a man made fiber made with two variants of the same basic chemical polymer.

Cotton - a natural cellulosic fiber that grows in the seed pod of the cotton plant; used primarily for apparel, household items, upholstery, and industrial products.

Crease resistant - fabrics that have been treated or constructed to improve their resistance to and recovery from wrinkling.

Crimp - describes the waviness of fibers, natural or manufactured.

Cross-dyeing - method of dyeing fabrics containing yarns with different affinities for different dyes.

Cross-linking - a chemical process for modifying the molecular structure of a fiber by forming a permanent bond or link.

Cross-section - the shape of an individual filament when cut at right angles to its axis.

Curl - general term used to describe the conformation of a textured filament yarn.

Denier - unit of weight indicating size of a fiber filament based on thickness (weight) in grams.

Dimensional stability - tendency of fabric to retain shape and size after wear, washing, and/or dry cleaning.

Edge crimping - a technique for texturing a filament yarn by heating it and drawing it over the edge of a sharp blade.

Elongation - deformation of a fiber due to stretching.

Fabric - any woven, knitted, plaited, braided, felted, or nonwoven material made of fibers or yarns.

False twist - yarn that has been twisted by a rotating spindle, set in a heat box, and then untwisted; it is called "false" twist, because the twist does not remain.

Feed yarns - yarns supplied to the throwster by the fiber producer, ready for texturing.

Fiber - any substance, natural or man made, composed of thread-like tissue capable of being spun, woven, or knitted.

Filament - an individual strand of continuous length from a silkworm cocoon, or a man-man synthetic strand of continuous length extruded through one hole in the spinneret.

Flax - a cellulosic fiber from the stem of a plant; fabric made from the flax fiber is called linen.

Gauge - the measurement of needles per inch in the knitting machine.

Gear crimping - a yarn texturing method in which the yarn passes through the meshing teeth of two heated gears.

Godet wheel - wheel over which newly spun man-made fiber is wound.

Hand - describes the touch, handle, or feel of fabrics.
Heat set - to fix or set a yarn in crimped or textured form by use of heat.
Hollow fibers - man made fibers spun through spinneret with pronounced "C" shaped holes.
Homo-polymer - a man made fiber that consists of one chemical polymer, as distinct from a co-polymer.
Hydrophobic - property of resisting the absorption of water-common to nylon, polyester, and other synthetics.
Knit-de-knit - a yarn texturing method.
Knit Machines:
    Jersey-flat - the basic knitting stitch; the jersey flatbed produces full-fashioned sweaters, as well as many other types.
    Purl-flat - the famous links-links machine, fine for making pullovers, cardigans, and the classic golf or tennis sweater.
    Jersey-circular - the industry "workhorse," producing underwear, seamless hosiery, piece goods, and other high volume apparel.
    Purl-circular - the high volume purl machine, producing cut and sewn sweaters, dress wear, and men's socks.
    Rib-circular - the classic example is the double knit; the fastest growing knitting machine category for men's wear.
    Tricot - a warp knit machine that produces fine vertical lines on the face and crosswise ribs on the back of fabric.
    Rib-flat - a complex machine, capable of stitch transfer and trimming cuffs and collars expertly.
    Raschel - a warp knit machine that can stitch or lay in yarn, create open or lacy fabrics, and rigid or stretchy fabrics.
Knitting - the process of making fabric by interlocking a series of loops of one or more yarns.
Loft - the properties of firmness, resilience, and bulk of a fiber batting, yarn, fabric, or other textile material.
Loom - a weaving machine for producing textiles by interlacing warp and filling yarns in perpendicular fashion.
Loop - the formation given to a yarn by the air jet method of texturing.
Luster - shine or sheen of fibers, yarns, or finished fabrics.
Man made cellulosic - fibers produced by modifying or transforming natural cellulosic polymers.
Manufactured Cellulosic Fibers - Rayon, acetate, triacetate are made from natural cellulosic raw materials such as wood pulp.
Modulus - a measure of the stress needed to stretch a fiber to the maximum extent from which it will return completely to its original shape.
Mono-filament - a single strand of man made fiber, continuous filament, or strand.
Multi-filament - yarns with many fine continuous filaments or strands; multi-filament yarns produce especially soft fabrics.

Natural fibers - the four natural fibers are cotton, flax, silk, and wool.
Nitrate rayon - the original rayon made by Chardonnet from nitrocellulose.
Non-woven - a material made of fibers in a web or mat held together by a bonding agent or by interlocking fibers as in felt.
Nub yarn - novelty yarn containing slubs, beads, or lumps intentionally introduced.
Pent-lobal - a man made fiber whose cross section has been modified from circular to five sided.
Permanent press - a pre-cured or post-cured process for producing a garment with a durably set shape and/or designed creases, which requires no ironing after normal laundering for its normal life.
Petroleum based fibers - fibers produced by the modification of polymers derived from petroleum.
Piece dye - a fabric dyed in the piece, after construction.
Producer of texture - a general term for yarns textured by the fiber producer at the source, rather than by a throwster.
Protein fiber - natural protein fibers, from animal hair or from the cocoon of a silkworm.
Raschel gauge - measured as needles per two inches.
4
Rayon - manufactured cellulosic fibers made from natural cellulosic raw materials such as wood pulp; rayon was used as a less expensive substitute for silk, prior to the man-made synthetic production of nylon.
Rayonne - French generic term for viscose rayon staple.
Recovery - the essential property of a stretch yarn.
Resilience - the property of a textile material to recover from a deformed state.
Scroop - crunching sound or sensation produced by rubbing together fibers having a high degree of inter-fiber, stick-slip friction.
Set - (as in "set" yarn) one that has been textured and heat set to remove stretch.
Set yarns - a false twist "set" yarn is really the same stretch yarn that has been given a second heat treatment.
Shrinkage - treatments to remove most of a fabric's tendency to shrink.
Solution-dyeing - process of "dyeing" fiber in the solution stage, before spinning, so that the filament is impregnated with pigment.
Silk - a natural protein fiber that comes from the cocoon of the silkworm and is used in apparel and some home furnishings.
Spindle - a rotating device with an eye through which yarn is threaded and then twisted.
Spinneret - a metallic cap or jet with microscopic holes in the surface through which spinning solutions are forced, becoming filaments as they emerge into a coagulating medium.
Staple - short or long lengths of fiber twisted together to make a yarn.

Stretch - (as in "stretch" yarn) one that has been false twisted and not set.
Stretching - (as applied to man-made fiber production) the stretching process gives the fiber greater strength after it has been extruded.
Stretch yarns - a filament yarn that has been twisted 50 or more turns per inch.
Stuffer box - a texturing method by which yarn is stuffed into one end of a heated box and emerges from the other end with a sawtooth crimp.
Synthetic fibers - fibers manufactured by chemical syntheses.
Tenacity - a measure of yarn strength, i. e., high tenacity denotes high strength.
Tensile strength - the comparative measurement of the strength of fibers, yarns, or fabrics.
Texture - derived from the Latin "texere," meaning to weave; now applied to crimped yarns and fabrics with surface interest.
Textured acetate - acetate filament yarn that has been bulked to produce a more lofty texture.
Thermoplastic - (as in "thermoplastic" fiber) a chemical fiber that becomes malleable when subjected to heat and whose form can be changed permanently by heat.
Throwster - derived from the old Anglo Saxon word "thrawan," meaning to twist or revolve.
Torque - the stretch factor tendency to curl, imparted to a yarn by false twisting.
Tow - a large number of continuous filaments drawn and collected in a loose, rope-like strand without definite twist; suitable for cutting into staple or flock.
Triacetate - made by modifying cellulose and resists more heat than acetate, it generally holds pleats, shape, and texture.
Tricot - French for warp-knit fabric.
Tricot gauge - measures in needles per one inch.
Tricot machine - designed to convert smooth, fine denier filament yarns into lingerie and bonding fabric.
Tri-lobal - a man made fiber whose cross section has been modified from circular to three sided.
Twist - the turning of fibers or yarns around their axes, expressed in number of turns per unit of length.
Twist factor - (also referred to as twist multiplier) the ratio of turns per inch to the square root of the yarn count.
Warp - one set of yarns that run along the length of the fabric.
Warp crimping - crimped into a multiple layer of ends arranged in warp formation.
Weaving - the process of manufacturing fabric by interlacing a series of warp yarns with filling yarns at right angles.
Weft - filling yarn that runs perpendicular to the warp.
Wet-laid - formed fabrics manufactured by using processes and equipment adapted from paper-making techniques.
Wicking - the dispersing or spreading of moisture or liquid through a given area, vertically or horizontally, capillary action in material; the quality of a yarn or fabric that enables it to draw moisture away from the body, thus making the fabric more comfortable to wear.
Wool - natural protein fiber that comes from animal hair and can be used in apparel, home furnishings, and for industrial uses.
Yarn - descriptive term for an assemblage of fibers or filaments, either manufactured or natural, twisted or laid together so as to form a continuous strand that can be used in weaving or knitting, or otherwise made into a textile material.

SELECTED BIBLIOGRAPHY

American Fabrics. New York: Doric Publishing Co., Issues 1970-1979.

American Fabrics. AF Encyclopedia of Textiles. 3rd ed. Englewood Cliffs, New Jersey: Prentice-Hall Inc., 1980.

Birell, V. L. *The Textile Arts: A Handbook of Fabric Structure & Design Processes*. New York: Harper & Row, 1959.

Brown, Patty, and Janett Rice. *Ready-to-Wear Apparel Analysis*. Upper Saddle River, New Jersey: Prentice-Hall, Inc., 1998.

Corbman, Dr. Bernard P. *Textiles: Fiber to Fabric*. New York: McGraw-Hill Inc., 1975.

Davis, Rebecca, and Carol Tuntland. *The Textiles Handbook*. Redondo Beach, California: Plycon Press, 1996.

Dictionary of Textile Terms, A. 12th ed. New York: Dan River Inc., 1976.

Garthe, Mary. *Fashion and Color*. Rockport, Massachusetts: Rockport Publishers, 1995.

Gioello, Debbie Ann. *Profiling Fabrics: Properties, Performance & Construction Techniques*. New York: Fairchild Publications, 1981.

Hall, David M. *Practical Fiber Identification*. 2nd ed. Auburn, Alabama: Auburn University Press, 1982.

Herald, Jacqueline. *Fashions of a Decade: The 1970s*. New York: Facts on File, Inc., 1992.

Hollen, Norma, Jane Saddler, Anna Langford, and Sara Kadolf. *Textiles*. 6th ed. New York, MacMillan Publishing Company, 1993.

Houck, Catherine. *The Fashion Encyclopedia*. New York: St. Martin's Press, 1982.

Humphries, Mary. *Fabric Glossary*. Upper Saddle River, New Jersey: Prentice Hall, 1996.

Jerde, Judith. *The Encyclopedia of Textiles*. New York: Facts on File, Inc., 1992.

Joseph, Marjory L. *Essentials of Textiles*. New York: Holt, Rinehart & Winston, 1976.

Kim, Charles J. *Textile Science An Outline*. Dubuque, Iowa: Kendall/Hunt Publishing Company, 1997.

Lauterburg, Lotti. *Fabric Printing*. New York: Reinhold Publishing Corp., 1959.

Mulvagh, Jane. *Vogue History of 20th Century Fashion*. London: Viking, 1988.

O'Hara, Georgina. *The Encyclopaedia of Fashion*. New York: Harry N. Abrams, Inc., Publishers, 1986.

Peacock, John. *The 1970s*. New York: Thames and Hudson, 1997.

Reichman, Charles. *Knitting Dictionary*. New York: National Knitted Outerwear Association, 1966.

Schnurnberger, Lynn. *Let There be Clothes*. New York: Workman Publishing, 1991.

St. James Fashion Encyclopedia: Survey of Style from 1945 to the Present. Detroit, Michigan: Visible Ink Press, 1997.

Tortora, Phyllis B. *Fairchild's Dictionary of Textiles*. Upper Saddle River, New Jersey: Prentice-Hall Inc., 1996.

Tortora, Phyllis B. *Understanding Textiles*. Upper Saddle River, New Jersey: Prentice-Hall Inc., 1996.

Wilson, Kax. *A History of Textiles*. Boulder, Colorado: West View Press, Inc., 1979.

Yeager, Jan. *Textiles for Residential and Commercial Interiors*. New York: Harper & Row, Publishers, 1988.

INDEX

Alamac Knitting Mills, 97
Ameritex, 21, 26, 80
Angle, 6
Avondale, 29, 96
Belgian Linen Association, 15, 45
Bloomcraft, 53
Bravo Knits, 10
Bronzini Sportswear, 9
C&V Fabrics, 39
Cannon Mills, 16
Caro-Knit, 9, 21, 36
Cohama, 8, 31
Collins & Aikman, 12
Compton, 76
Cone Mills, 7, 11, 40, 70, 73
Continella Industries, 49
Continella Textile Corp., 49
Couleur, 102
Covington, 69
Cranston, 58, 77
Crantex, 60, 102
Crompton, 6, 38, 87, 89
Crompton-Richmond, 69
Cyanamid, 38
Cyrus Clark, 70, 102
Dan River Mills, 18, 27, 29, 36, 38, 39, 40
Deering Milliken, 44-45, 54
Domestic Fabrics Corp., 22
Earl-Glo Fabrics, 14
Everfast, 75
F.C.I, 43
Galey & Lord, 7, 26, 39, 86
Graniteville, 56
Greeff, 52, 75
Hatema-Texoprint, 36-37, 44, 49, 72, 92
Held Fabrics, 9, 11
J.B.J. Fabrics Inc., 68
J.P. Stevens & Co., 27, 86
Kenmill, 104
Knoll Textiles, 32
Lowenstein, 16, 97
M. Lowenstein & Sons, 42
Mantero, 68
Milliken, 47
New York Trim n' Fashion, 23
Novelty Textiles, 9
Onondaga, 10, 71
Orinoka Mills, 11, 21, 47
Quality Knitting Mills, 93
R.B.S., 40
Riverdale, 48
Riverdale Drapery Fabrics, 36
Russell Mills, 15, 20
Schumacher, 55
Schwartz-Liebman Textiles, 93
Seta-Sole International Ltd., 54
Silkorea, 12, 23, 53
Singer, 22
Springmaid, 19, 41
Spring Mills, 43, 55
Sopra, 42
Stevcoknit Fabric Company, 20, 94
Stevens Hockanum, 19
Stroheim & Romann, 81
Sulzer Bros., 17
Sunrise Drapery Co., 15
Swift Textiles, 20-21, 27-28, 30
Transprints International, 66, 75, 80
Tromer Fabrics, 30
Viisco, 39, 49, 86
Wamsutta Fabrics, 8, 37, 60, 87
Waverly Fabrics, 29
Wedgwood, 17
West Point Pepperell, 52
William Heller, Inc., 87
Wulschleger, 93
Zegna, 8